WEIRD WORLD OF ANIMALS

SURPRISING FLYING ANIMALS

BY MARY GRIFFIN

Gareth Stevens PUBLISHING

Please visit our website, www.garethstevens.com. For a free color catalog of all our high-quality books, call toll free 1-800-542-2595 or fax 1-877-542-2596.

Names: Griffin, Mary, 1978-.
Title: Surprising flying animals / Mary Griffin.
Description: Buffalo, NY : Gareth Stevens Publishing, 2026. | Series: Weird world of animals | Includes glossary and index.
Identifiers: ISBN 9781482471861 (pbk.) | ISBN 9781482471878 (library bound) | ISBN 9781482471885 (ebook)
Subjects: LCSH: Animal flight--Juvenile literature. | Animal locomotion--Juvenile literature.
Classification: LCC QP311.G75 2026 | DDC 591.5'7--dc23

Published in 2026 by
Gareth Stevens Publishing
2544 Clinton Street
Buffalo, NY 14224

Cover Designer: Nicholas Switalski
Interior Designer: Rachel Rising
Editor: Therese Shea

Photo credits: Cover, p. 1 Travis Potter/Shutterstock.com; Cover, pp. 1, 3-24 8H/Shutterstock.com; p. 4 Nejron Photo/Shutterstock.com; p. 5 takmat71/Shutterstock.com; p. 7 Dr. Meet Poddar/Shutterstock.com; p. 7 dwi putra stock/Shutterstock.com; p. 9 Mohamad alias/Shutterstock.com; p. 11 Lauren Suryanata/Shutterstock.com; p. 13 Brent Barnes/Shutterstock.com; p. 14 L.T. Southall/Shutterstock.com; p. 15 Agami Photo Agency/Shutterstock.com; p. 17 Roberto Dani/Shutterstock.com; p. 19 Shpatak/Shutterstock.com; p. 21 Maqdis Studio/Shutterstock.com.

Printed in the United States of America

CPSIA compliance information: Batch #CSGS26: For further information contact Gareth Stevens, at 1-800-542-2595.

CONTENTS

*Words in the glossary appear in **bold** type the first time they are used in the text.*

Awesome Adaptation

Would you like to be able to fly? Imagine being able to get from place to place by air. It would certainly help if you lost a kite in a tree—or if you were stuck in a traffic jam!

Flying is an animal adaptation. An adaptation is a body feature or **behavior** that a living thing has **evolved** to survive. Flying is certainly a handy adaptation for birds and bats, for example. But there are other, more surprising, creatures that get from place to place by air.

Some birds lost their ability to fly over time. These include penguins. Penguins evolved to become better swimmers than fliers because they find their food in water.

That's WEIRD!

Adaptations can keep changing. Birds in the ratite bird group—including ostriches—may have lost and gained the ability to fly as many as five times!

FLYING SQUIRRELS

Have you heard of flying squirrels? Some squirrels seem to fly as they fearlessly jump from tree branch to branch. But one group of squirrels really do have adaptations to increase their airtime. In fact, they're called flying squirrels. They don't really fly—they **glide**—but they can get *really* far.

The giant flying squirrel of Asia is the biggest. It has extra flaps of skin and muscle between its legs. It can leap into the air, stretch out these flaps, and glide as far as 1,500 feet (457 m)!

Flying squirrels' extra flaps of skin work a bit like a *glider* or a *parachute*.

That's WEIRD!

Flying squirrels have long, flat tails that guide them as they move through the air.

Parachute Frogs

Squirrels aren't the only tree-dwelling animals that have a "flying" adaptation. Parachute frogs, also called Wallace's flying frogs, live in some rainforests of Southeast Asia. They use large webbed feet to glide from tree to tree or to the ground. These small **amphibians** may glide 50 feet (15 m) at a time!

Parachute frogs' special feet also help them stick to tree trunks and branches. They get around like this to either escape predators or catch **prey**.

Parachute frogs grow to be about 4 inches (10 cm) long.

That's WEIRD!

Parachute frogs only come down from trees to **mate** and lay eggs.

COLUGOS

There are two species, or kinds, of gliding **mammals** called colugos. They live in Southeast Asia, including a few of the Philippine Islands. They "fly" much like flying squirrels, with flaps of skin stretched out between their limbs. **Membranes** also connect their back legs and tail. Colugos can glide more than 200 feet (61 m) from one tree to the next.

Colugos are sometimes called flying **lemurs**, but they're not lemurs. They just look a bit like lemurs because they have big eyes. They're also nocturnal, or more active at night, like lemurs are.

Colugos eat leaves, flowers, tree sap, and fruit, which they can find in their tree homes. If they can't, they glide to another tree!

That's WEIRD!

Colugo babies hold on to their mothers as the mothers glide from tree to tree.

FLYING FISH

About 40 species of fish in the ocean are called flying fish. No, they can't really fly, but some look like they can! Most flying fish are small, less than 18 inches (46 cm) long. They have fins that look a bit like wings when they're in the air and above the water. They sometimes have two fins and sometimes four.

Why do they fly? Flying fish are usually trying to get away from predators in the water.

Flying fish are often found in warm ocean waters.

That's WEIRD!

Flying fish can get so high into the air that the end up on the decks of ships!

Here's how flying fish "fly". They start out underwater. They begin to increase their swimming speed, keeping their fins close to their body and getting faster and faster. Then they come up into the air and spread their fins. Their tail may beat against the water to push them up more.

Flying fish may glide this way up to 600 feet (183 m) at a time. They can travel at a speed of more than 10 miles (16 km) per hour.

This pictures of a flying fish shows that it uses its tail to push itself along, above water.

FRESHWATER FLIERS

Not all flying fish live in salt water. The freshwater butterflyfish of Africa lives in freshwater ponds and rivers. It uses its adaptation of nearly flying to catch its prey.

It can sense **insects** touching the surface of the water. When it does, the butterflyfish propels, or pushes, itself with its fins into the air and glides to the exact location of the insect! The freshwater butterflyfish may also use its jumping and gliding abilities to get away from predators.

Butterflyfish are named for their fins, which look a bit like a butterfly's wings.

That's WEIRD!

African freshwater butterflyfish can breathe air for a short time when they come out of the water!

JAPANESE FLYING SQUID

Did you ever run through water? Water really slows you down. The creature called the Japanese flying squid has evolved an adaptation to move through air so it can move more quickly. It doesn't have skin flaps or fins, so how does it do this?

The Japanese flying squid sucks water into its mantle, the main part of its body. Then, it pushes the water out with great force through a tube-shaped body part called a siphon. The powerful push can lift the squid out of water.

The Japanese flying squid lives in the northern Pacific Ocean. These squid aren't large creatures, only reaching about 20 inches (50 cm) long.

That's WEIRD!

The Japanese flying squid can move about 36 feet (11 m) per second!

MANY MORE SURPRISES

These animals aren't the only flying or gliding creatures of nature. The draco lizard glides around in Southeast Asian forests to find food, escape predators, and show off for mates. The common flying gecko is an Asian lizard with skin flaps, a flat tail, and webbed feet to help it glide. And there's still others!

These animals' body parts and behaviors work together to help them move through air, just as other animals' adaptations help them fly. Flying and gliding are key adaptations in the wild!

These animals use their special adaptations for different reasons—all help keep them alive!

Other Surprising Creatures That *Almost* Fly

GREATER GLIDERS the largest of the gliding **marsupials** in Australia, they have big, furry ears

PARADISE TREE SNAKES Southeast Asian snakes that can make their body flat to glide in the air

CRAB SPIDERS spiders that live around the world and balloon, or seem to fly with the help of a long silk thread

MOBULA RAYS sea creatures that live in warm seas and leap into the air with a flat, diamond-shaped body

That's WEIRD!

Draco lizards can tuck their "wings" into their ribs!

GLOSSARY

amphibian: An animal that spends time on land but must have babies and grow into an adult in water.

behavior: The way an animal acts.

evolve: To grow and change over time.

glide: To move in a smooth and graceful way.

glider: A light aircraft made to fly long periods without using an engine.

insect: A small, often winged, animal with six legs and three main body parts.

lemur: A small animal from Madagascar with thick fur and a long tail that lives in trees and is active at night.

mammal: A warm-blooded animal that has a backbone and hair, breathes air, and feeds milk to its young.

marsupial: A mammal whose young develop in a pouch on the mother's belly.

mate: To come together to make babies. One of two animals that come together to produce babies.

membrane: A soft, thin tissue of the body.

parachute: A specially shaped piece of cloth that collects air to slow something down.

prey: An animal that is hunted by other animals for food.

FOR MORE INFORMATION

Books

Hecht, Tracey. *Unique Adaptations of Nighttime Animals*. New York, NY: Fabled Films Press, 2022.

Pringle, Laurence. *The Secret Life of the Flying Squirrel*. New York, NY: Astra Young Readers, 2023.

Websites

Flying Fish
www.nationalgeographic.com/animals/fish/facts/flying-fish
Learn more about flying fish from National Geographic.

Northern Flying Squirrel
www.biokids.umich.edu/critters/Glaucomys_sabrinus/
Read about northern flying squirrels—which may live near you!

Publisher's note to educators and parents: Our editors have carefully reviewed these websites to ensure that they are suitable for students. Many websites change frequently, however, and we cannot guarantee that a site's future contents will continue to meet our high standards of quality and educational value. Be advised that students should be closely supervised whenever they access the internet.

INDEX